Contents

Introducing more great titles in The Unexplained series.

When it comes to the world of the supernatural and the paranormal, there's a lot to talk about. Together with *Alien Encounters*, these are the other brilliant books in this brand new series hot off the Hamlyn production line.

THE UNCANNY

Funny goings-on in mind, body and spirit

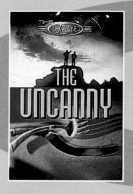

MYSTERIOUS PLACES

Strange happenings in remote regions

HAUNTINGS

Spooky travels through time and space

Alien Encounters

Have you ever seen an Unidentified Flying Object, a "UFO"? If you have, then don't worry — you're not a crank! Thousands of people have reported seeing one. But that's not the same thing as identifying an alien spacecraft. About 95 per cent of "sightings" soon turn out to be IFOs — *Identified* Flying Objects.

The sky is a busy place, and it's all too easy to mistake planes, balloons, comets, stars and even the Moon for "flying saucers". But what makes UFOs fascinating is the 5 per cent of reported sightings that do remain genuine mysteries.

Alien Encounters highlights five classic UFO incidents — and examines the vast range of theories put forward to explain them. When you read about UFOs, you need to start with an open mind. Nonsense or common sense? The Unexplained series lets YOU decide what to believe...

THE SAUCER SCARE

A civilian pilot, Kenneth Arnold, was using his plane to search a remote area of the Cascade Mountains for a missing US Marine transport aircraft. The Government had offered a $5,000 reward to anyone who could locate the wreckage. As an experienced pilot, Arnold hoped that the reward would be his before nightfall.

DATE:

24 June 1947

TIME:

3 p.m.

PLACE:

Washington State,
United States
of America

Carefully scanning the ground, he banked his aircraft in a sweeping turn over the town of Mineral. Flying at 2,750 metres, he could see the snow-capped heights of Mount Rainier glistening in the distance. His tiny, single-engined Callair was perfect for searching the rough terrain below. And if there was a problem, he could land it on a "postage stamp".

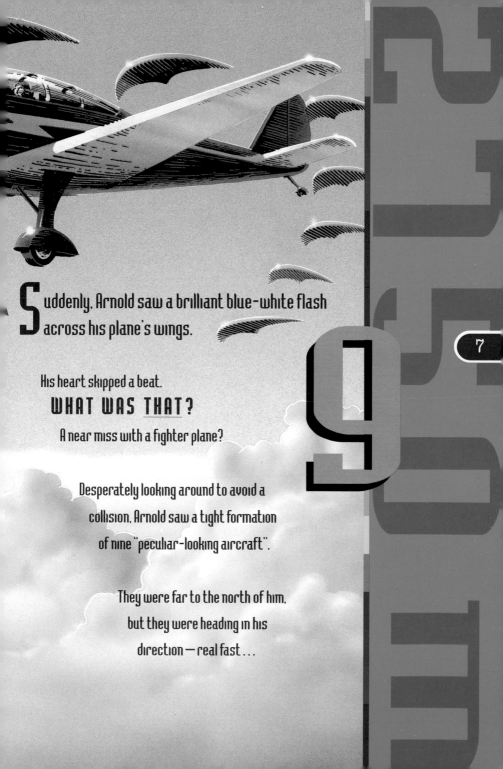

Suddenly, Arnold saw a brilliant blue-white flash across his plane's wings.

His heart skipped a beat.

WHAT WAS <u>THAT</u>?

A near miss with a fighter plane?

Desperately looking around to avoid a collision, Arnold saw a tight formation of nine "peculiar-looking aircraft".

They were far to the north of him, but they were heading in his direction — real fast...

They were approaching Mount Rainier very rapidly and I assumed they were jet planes. Two or three of them would change their course every few seconds...just enough for the Sun to strike them at an angle.

8

00km/h
00km/h
00km/h
00km/h
00k

> " They didn't fly like any aircraft I'd seen before. Maybe it would be best to describe their flight characteristics as being similar to a flock of geese, in a rather diagonal chain-like line. They fluttered and sailed, tipping their wings alternately, and emitting those bright blue-white flashes from the extremely highly-polished surfaces of their wings. "

00km

00km/h

9

00km/h   Glancing at the dashboard clock, Arnold noted the time at a little before 3 p.m.

00k 5 /h 0   0   h   o   u   r   s

00km/h   He tracked the objects for the next three and a half minutes. Watching them zoom between the peaks of Mount Baker and Mount Rainier, he estimated their speed at 2,000 kilometres per hour. This was incredible — almost three times faster than any jet of that time!

00km/h

00km/h

THE PILOT WAS AMAZED

He was sure the United States didn't have such advanced aircraft. But what about the Soviet Union? Had he just

00km/   witnessed secret Soviet planes crossing the Pacific Ocean from Siberia

00km   to test out the US air defences? When Arnold landed in Pendleton,

00km/h   Oregon, he wanted to talk to the Federal Bureau of Investigation, the famous FBI. But the local office was closed, so ...

... INSTEAD HE SPOKE TO THE PRESS.

Arnold told journalist Bill Bequette that his mystery planes flew like "speedboats in rough water". Thinking again, he added: "They flew like a saucer would if you skipped it across water." Bill knew he had found a sensational story and wrote down the phrase "flying saucer", even though this was not what Arnold had said. Within a few days Arnold's "saucer" story had made headlines across the world.

The United States Air Force (USAF) was concerned enough to set up its first official investigation, Project Sign, in 1948.

THIS ADMITTED THAT 20 PER CENT OF UFO SIGHTINGS WERE DIFFICULT TO EXPLAIN.

UFOs

People have been seeing mysterious objects in the skies for thousands of years. After the poor reporting of the Arnold story these were often called **"flying saucers"**. In the 1960s the more scientific term Unidentified Flying Objects, or UFOs for short, came into common use.

Kenneth Arnold was, to all outward appearances, a pillar of society. He had been a successful businessman at a young age, and he also held a public post of responsibility — as an acting deputy US Marshal.

Washing...

Seattle Post Intelligence

25 June 1947

MYSTERY DISK HURTLING ACROSS SKY

While searching for a missing US Marine transport aircraft in the Cascade mountains Kenneth ...saw a strange sight...

ERN STAR

SAUCER

25th June 1947

THE PRICE OF FAME

In later years, Arnold reported several more sightings — leading some people to believe that he had always been seeking publicity. At first, Arnold was treated as an honest witness. But soon many commentators were calling him a fraud or a liar. In 1984 he looked back on the events of 1947 and wrote:

> "Nameless, faceless people made
> fun of me. I was considered a fraud.
> I loved my country. I was very naive.
> I was the unfortunate goat who first
> reported them [UFOs]."

11

Spot Your Flying Saucer

FLAT BOTTOM, CONCAVE				
CONICAL DISCS				
COIN-LIKE DISC WITH VARIATIONS				
BI-CONVEX LENTICULAR DISCS				
BI-CONVEX DOMED DISCS				
CYLINDRICAL (CIGARS)				
ANGULAR: CUBES, TETRAHEDRAS, CRESCENTS				

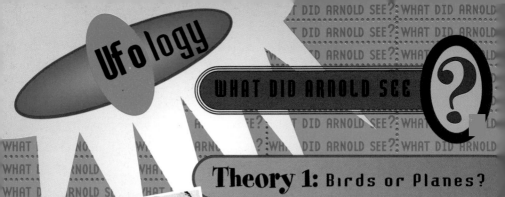

Ufology

WHAT DID ARNOLD SEE?

Theory 1: Birds or Planes?

Drawings he made after the event showed bird-like or crescent-shaped objects with swept-back wings and no tail section. Had he watched a flock of birds and badly misjudged their speed and distance? There were also several USAF bases in the area, but no planes were reported as flying that afternoon.

12

Theory 2: A Guided Missile

A local pilot suggested Arnold had seen guided missiles. These were often test-fired from a base at Moses Lake near the Cascade Mountains. But missiles travel in a straight line, not swooping and soaring like Arnold's mystery aircraft.

Theory 3: Unusual Clouds

Meteorologists point out a third possibility — the atmosphere playing tricks. If a layer of cold air moves close to the Earth's surface this can produce strangely shaped clouds. Arnold may have mistaken these for UFOs.

Whatever the truth about Arnold's

Air Force Alert

On 28 June 1948, the USAF gave alarming support to Arnold's UFO sighting. A pilot in a P-51 Mustang fighter over Lake Mead in Nevada saw five or six circular objects flying in close formation...

story, public reaction to it was amazing.

That same night, four USAF officers saw a brilliant light zig-zagging across the sky before doing an impossibly sharp turn and zooming off at high speed.

Over the coming years, thousands

13

They knew that no US plane was able to perform brilliant aerobatics like that.

# of people reported sightings of

The Cold War Connection

During World War 2, the Soviet Union and the USA were allies against Nazi Germany. When the war ended these two superpowers became rivals. This rivalry became known as the "Cold War" — and by 1947 the US was beginning to think of the Soviet Union as its enemy.

strange saucer-shaped objects.

The possible threat of Soviet secret weapons worried the US government. After Arnold's report, US camera planes patrolled the skies looking for secret enemy aircraft.

THE MODERN AGE OF UFOs HAD BEGUN.

Nobody knows where the invading force came from. It was first seen as it flashed across the dawn sky of western China and plunged into the Earth's atmosphere. Racing towards the planet, it hit denser layers of air over the Gobi Desert and became a glowing fireball...

In seconds it crossed over Mongolia, burning brighter than the morning sun. A powerful shock wave shook the ground, smashing trees and animals like specks of dust.

At **7.17 a.m.** the "UFO" struck the valley of the Tunguska River in Siberia. There was a massive explosion. A giant pillar of fire blazed into the air and could be seen for hundreds of kilometres. The noise was so great that herdsmen 50 kilometres away were deafened. An area the size of a large city was laid waste, and herds of reindeer were incinerated as they fled their summer pastures.

DATE:

30 June 1908

PLACE:

Tunguska region,
Central Siberia,
Russia

A searing heatwave ripped outwards from the blast, scorching ancient conifers and igniting fires that burned for days. Sixty kilometres away the frightened people of Vanavara turned their faces from the burning wind. Seconds later the blast wave reached their village, shattering windows and roofs. A farmer sitting on his porch described it:

There was an immense flash of light. There was so much heat my shirt almost burned off my back. I blacked out for a moment and when I came to I heard a noise that nearly shook the house off its foundations.

five hundred kilometres from the explosion the Trans-Siberian Express was nearing the station at Kansk. The train jarred wildly on the tracks and the driver saw the railway lines shaking. Startled, he brought his locomotive to a shuddering stop.

A Harsh Land

Siberia is a vast area of Russia. It takes five days to cross on the famous Trans-Siberian Express train. Much of it is remote and merciless. Huge forests are locked in ice in winter, when temperatures frequently drop below -40° C. But in the brief summer the Tungus tribespeople can find rich grazing for their reindeer herds.

A
NEW YORK TIMES
REPORTER, IN ENGLAND,
WROTE:

Earthquake detectors in the city of Irkutsk, 750 kilometres away,
jumped into life as they recorded powerful tremors.
The vibrations circled the world, registering strong
shocks as far away as Germany and America.
Dust and dirt were hurled 20 kilometres high
and fell back as sinister black rain.
Huge clouds blanketed northern
Russia and parts of Europe
for days. The effect
was eerie.

17

"The northern night sky became light blue, as if the dawn
were breaking, and the clouds were touched with pink.
Police headquarters was rung up by several people who
believed that a big fire was raging in the north of London."

AN AWESOME FORCE DESTROYED THE PEACE OF THAT REMOTE SIBERIAN SUMMER ...

UFology

WHAT CAUSED THE SIBERIAN EXPLOSION ?

WHAT CAUSED THE SIBERIAN EXPLOSION WHAT CAUSED THE SIBERIAN EXPLOSION WHAT CAUSED THE SIBERIAN EXPLOSION

Theory 1: A Meteorite or Asteroid

In 1927 Leonid Kulik led the first scientific expedition to study the events at Tunguska. He was shocked by the scale of the devastation and believed a large meteorite, around 30 metres in diameter, had struck the area. He found flattened or burned trees fanning out over hundreds of square kilometres from the centre of the explosion — but there was no crater. He guessed that the meteorite had broken up into tiny pieces.

18

Theory 2: A Comet

Some scientists have suggested that a small comet, only 150 metres in diameter, could have hit Tunguska. Comets are made largely from ice, with a small solid core. A comet plunging through the Earth's atmosphere would have reached a temperature of millions of degrees — enough to explode and vaporise.

Theory 3:
An Alien Spacecraft

In 1945 the USA exploded atomic bombs on Hiroshima and Nagasaki in Japan. The Russian writer Alexander Kasantsev noticed that photographs of the damage in Siberia and Japan were similar. As atomic bombs did not exist in 1908, Kasantsev suggested that an alien spacecraft, powered by atomic motors, had exploded over Tunguska. Pillars of fire, dust clouds and black rain are all elements of nuclear explosions ...

19

CAN YOU SPOT A METEORITE? Look out for streaks of light, called shooting stars, in the sky on clear nights. They are meteorites burning up as they enter the Earth's atmosphere.

The Doomsday Strike

An asteroid 11 kilometres across hit the Earth about 65 million years ago in the Gulf of Mexico. Many scientists believe this changed the world's climate and wiped out **the dinosaurs.**

20

DANGEROUS SPACE JUNK

METEORITES AND ASTEROIDS ARE PIECES OF ROCK DRIFTING IN SPACE.

THE EARTH IS UNDER CONSTANT BOMBARDMENT FROM THESE DEADLY ROCKS →

I n November 1996 an international conference in London met to discuss the risk of humanity becoming extinct — if the world was hit by a big enough asteroid. They heard that an asteroid only 50 metres across would be enough to wipe out a country the size of England.

TO HELP AVOID THIS DISASTER, SCIENTISTS BELIEVE THEY COULD:

- Track thousands of objects in orbit near Earth
- Predict the time and place of an impact
- Knock the object off course with a nuclear missile. Expensive — and very dangerous!

21

England

THEY ARE THE REMAINS OF PLANETS AND STARS.

BUT LUCKILY MOST ARE SMALL AND BURN UP HARMLESSLY IN THE ATMOSPHERE.

CAUGHT ON TAPE

The radio crackled on. Control sat up to attention as a distressed voice came over the airwaves . . .

7:07 P.M.

DSJ: MELBOURNE CONTROL... IS THERE ANY AIR FORCE ACTIVITY IN THIS AREA?

CONTROL: DELTA SIERRA JULIET... NO KNOWN AIRCRAFT IN THE VICINITY.

7:08 P.M.

DSJ: IT'S APPROACHING FROM DUE EAST TOWARDS ME...IT SEEMS TO ME THAT HE'S PLAYING SOME SORT OF GAME. HE'S FLYING OVER ME – TWO, THREE TIMES – AT SPEEDS I CAN'T IDENTIFY. IT IS NOT AN AIRCRAFT! IT IS...A LONG SHAPE... CANNOT IDENTIFY IT MORE...IT HAS SUCH SPEED! IT'S RIGHT BEFORE ME NOW...

22

DATE:

21 October 1978

TIME:

7.07 p.m.

PLACE:

Bass Strait, Australia

CONTROL: DELTA SIERRA JULIET ...AND HOW LARGE WOULD THE – ER – OBJECT BE?

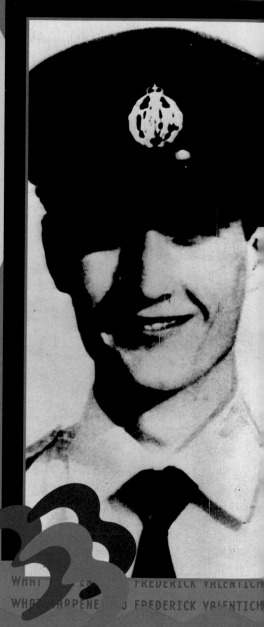

Delta Sierra Juliet (DSJ IN THE REPORT) was the radio call sign belonging to Frederick Valentich, a 20-year-old civilian pilot. He is pictured here in his Air Cadet Corps uniform. On 21 October 1978 he had rented a blue and white Cessna 182 to fly from Moorabbin Airport in Melbourne to King Island, near Tasmania. Forty-five minutes after take-off, at 7.07 p.m., he radioed in — and you have just read the recorded report. He was flying at 1,525 metres, closely followed by an aircraft he could not identify.

Steve Robey, the flight controller (CONTROL IN THE REPORT) listened carefully in Melbourne. But he could hardly believe what he was hearing. The last sound he heard was "a long metallic noise".

24

WHAT HAPPENED TO FREDERICK VALENTICH
WHAT HAPPENED TO FREDERICK VALENTICH
What happened to Frederick Valentic
What happen d to Fr d rick Val ntic
What happened to Frederick Valentich?

THREE MAIN EXPLANATIONS HAVE BEEN PUT FORWARD:

Theory 1: A UFO Attack

The tape of Valentich's radio messages seems to offer powerful evidence of hostile UFOs. On the very evening of Valentich's disappearance, a bank manager and his wife were out for a drive near Melbourne when they spotted a green starfish-shaped object over the sea. It was in the same area that Valentich's plane had gone missing.

25

Theory 2: An Accident

The official air accident inquiry stated that Valentich was "missing, presumed dead". The plane has never been found. However, pilot error is a likely cause. This was Valentich's first solo night flight, and his radio message appears to show he was very confused.

Theory 3: An Elaborate Hoax

Valentich was a UFO fan, and he even took a UFO scrapbook with him on the day he vanished. Perhaps Valentich planned his own disappearance to become famous. One later story claimed he was alive and well, working at a petrol station in Tasmania. But police checks failed to confirm this. His own family couldn't think of any reason why the young pilot might have wanted to fake his death.

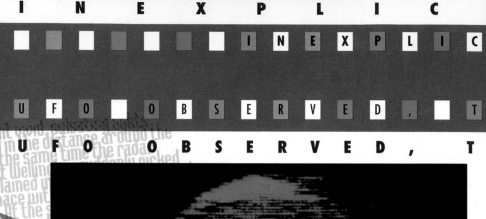

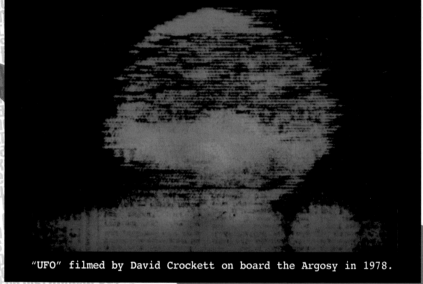

"UFO" filmed by David Crockett on board the Argosy in 1978.

W eeks after the Valentich incident the crews of two aircraft spotted bright unidentified lights between Wellington and Christchurch, in New Zealand. Ten days later, on 31 December 1978, a television team went up in an Argosy cargo plane to investigate.

Around midnight vivid, pulsating lights flickered in the distance around the plane. At the same time the radar station at Wellington suddenly picked up unexplained images. One blip seemed to keep pace with the Argosy for a few minutes. Witnesses on board saw a flashing light running parallel to them. The Argosy's own radar showed that one light came as close as 15 kilometres. A TV crewman reported this as having a . . .

"BRIGHTLY-LIT BOTTOM AND A TRANSPARENT SPHERE ON TOP."

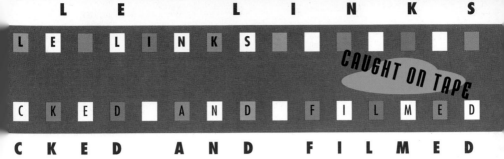

Studying the Evidence

The TV cameraman shot over 23,000 frames of 16mm colour film. This was studied by Dr Bruce Maccabee, an optical physicist working for the United States Navy. One sequence was "enhanced" by computer and showed a bell-shaped image that was brighter at the bottom — just as the cameraman had described. Maccabee also calculated that the object had indeed come within 15 kilometres of the plane, and that it measured 40 to 70 metres across. He estimated the power of its light at around 100,000 watts. He showed the film to other scientists, but none of them could explain what had caused the dancing lights. They really were dealing with "unidentified flying objects".

27

The UFO light analysed by Dr Maccabee was very strong — as brilliant as the beam from a lighthouse or a powerful *searchlight.*

Australia's "Bermuda Triangle"

The Bass Strait has been called the Australian "Bermuda Triangle" — a place where planes and ships have vanished without trace. During World War 2, some 17 aircraft were lost in these waters, even though there was no fighting in the area. Today the strait is one of Australia's UFO "hotspots".

CAMERA-SHY UFOs?

In 1977 the American newspaper *The National Enquirer* offered a million-dollar reward for proof that UFOs came from outer space. This was never won. Yet a good, clear photograph would have convinced the newspaper. Most photos turn out to be hoaxes, flaws in the camera or the film, or mistakes. The few *really* puzzling shots, like those from the New Zealand TV crew, are taken from too far away to prove anything.

Snap a Saucer!

If you're lucky enough to see a UFO, don't panic. Just get a picture. Grab any camera or camcorder (with film in!) and shoot away. You could be a celebrity on TV news right round the world — like George Adamski (*right*). And you don't have to be an experienced UFO spotter, either: the famous picture on the left was snapped by an English schoolboy called Stephen Pratt in 1966.

29

Remember to hold the camera steady — even if you're trembling with excitement — to help keep the image in focus. Happy hunting!

Most picture sightings we have of UFOs are poor quality because they are shot by amateurs. The professionals reckon that the best tool for the job is a 35mm single lens reflex camera (SLR), with a zoom lens. But they're not cheap.

For night patrol you really need a tripod and a cable release, which let you take long exposures without touching the camera and risking a blurred image.

If you're serious, load up with 36-exposure high speed (ASA 400) film and don't rely on batteries — some UFOs seem to be able to drain electrical power!

UFO ALERT!

"I had the feeling we were just chasing something that was playing with us. It had complete control."

The Belgian Gendarmerie (police) were being swamped with telephone calls. Hundreds of alarmed witnesses reported a huge triangular craft hovering over the countryside. The shape was easy to pick out in the clear night sky.

Red, green and amber lights glowed at each point of the triangle. Some people spoke of it moving at only 40-5 kilometres per hour, slower than a ca Most of the sightings came from the Wavre area, to the south of Brussels

DATE:
30 March 1990

TIME:
11 p.m.

PLACE:
South Belgium

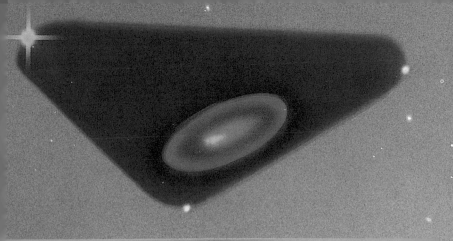

Other people claimed that it shot away at incredible speeds — from a standing start to thousands of kilometres an hour in just a few seconds.

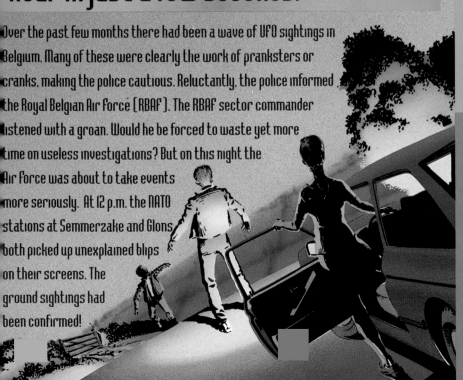

Over the past few months there had been a wave of UFO sightings in Belgium. Many of these were clearly the work of pranksters or cranks, making the police cautious. Reluctantly, the police informed the Royal Belgian Air Force (RBAF). The RBAF sector commander listened with a groan. Would he be forced to waste yet more time on useless investigations? But on this night the Air Force was about to take events more seriously. At 12 p.m. the NATO stations at Semmerzake and Glons both picked up unexplained blips on their screens. The ground sightings had been confirmed!

At 2.05 a.m. the RBAF emergency force known as the Quick Reaction Alert was scrambled. Two F-16 interceptor fighters always stood ready in case of any trouble, and they were airborne within minutes. As the fighters closed in, their radars locked on to the target. Their pilots were confident they would soon catch the intruder. But they were in for a shock, as one of them recalled:

> We eventually got it...the radar picture gives the altitude, the speed, the direction...what the UFO is actually doing. At the time we picked it up it was just travelling about 90 km/h. It was about 25 kilometres away when it seemed to say 'Well now, that's enough'. The airspeed went straight to Mach 8...Mach 9...Mach 10....incredible! And the altitude — it went from 1,500 metres straight up to 23,000 metres in just a split second!

Each time the fighters closed in, the intruder climbed or dropped away at astonishing speeds. The futile chase lasted almost an hour — until the UFO crash-dived to below 200 metres. This was too low for the radars to track it. The game was over. At the same time, ground observers watched the lights of the visiting UFO fade quickly away.

UFO ALERT!

33

Theory 1: A UFO Intruder

The Belgian Government investigated the incident, but decided there was not enough evidence to tell what the object was. The Minister of Defence admitted that an unidentified craft *had* flown over Belgium that night, changing its height and speed. One writer, Derek Sheffield, claims that the object was seen by 13,500 people from the ground and tracked by five different radar stations. He believes the same UFO flew over Britain.

34

Theory 2: American Cover-up

In the early 1990s the United States Air Force was testing secret aircraft. The B-2 Stealth bomber and the F-117A Stealth fighter were designed to fool radars. Both have a creepy triangular shape. Both are black and look menacing.

The USAF denied the plane had flown over Europe. However, the plane was based in USAF airfields in Britain at this time. It is likely the Americans *did* give their Stealth planes trial runs against the defences of friendly countries — before they had to be used in a real war.

Theory 3: False Findings

Belgian investigators think that evidence for the UFO scare may not be as strong as it appears.

● Witnesses reported seeing different things — flying rectangles, diamonds and boomerangs. The night was clear, so many sightings were likely to be planets or stars.

● The red, green and amber lights on the UFO are the same colours as the safety lights on ordinary aircraft.

● The radar of one F-16 fighter chasing the UFO "locked on" to the second fighter, giving unusual displays. All ground-based radars were affected by odd conditions in the atmosphere that day — and gave false trackings.

36

HELICOPTER V UFO

On the night of 18 October 1973, a US Army Bell UH-1 helicopter was flying at 760 metres near Mansfield, Ohio. This type of helicopter was nicknamed Huey. Captain Lawrence Coyne was in command of a crew of four. Just after 11 p.m. a red light zoomed in on a collision course with the Huey. Thinking it was a jet fighter-plane, Coyne power-dived to get out of its way.

In spite of his efforts, his altimeter showed he had climbed to 1,050 metres. Impossibly, the object stopped and hovered over the helpless Huey, bathing the 'copter in a strong green light. After a few moments it moved slowly away. Coyne desperately tried to call for back-up, but the radio stayed dead until ten minutes after the incident.

THE DARKSIDERS

In the hit TV series *The X-Files*, FBI agents Mulder and Scully discovered that aliens were working with the American Government. A group of gloomy UFO writers nicknamed "The Darksiders" believe that several crashed alien spacecraft have been found and scientists are studying their advanced technology. The public has been kept deliberately ignorant of this. Worse still, they believe world governments are working together with aliens to...

CONTROL THE REST OF HUMANITY!

UFO IDENTIFIED

*T*he storm was unusually heavy, and rain fell in torrents that bounced on the pavements. Thunder crashed from the heavy clouds while lightning flashes tore through the dull evening sky. A number of unlucky houses were damaged by strikes.

The ferocious power of the lightning knocked down chimneys and ripped off tiles.

Thankfully safe at home, a young woman was working in the kitchen. She was blissfully unaware that she was about to become an unwilling witness to a scientific riddle — the strange phenomenon of

BALL LIGHTNING.

DATE:

8 August 1973

TIME:

7.45 p.m.

PLACE:

Smethwick,
near Birmingham,
England

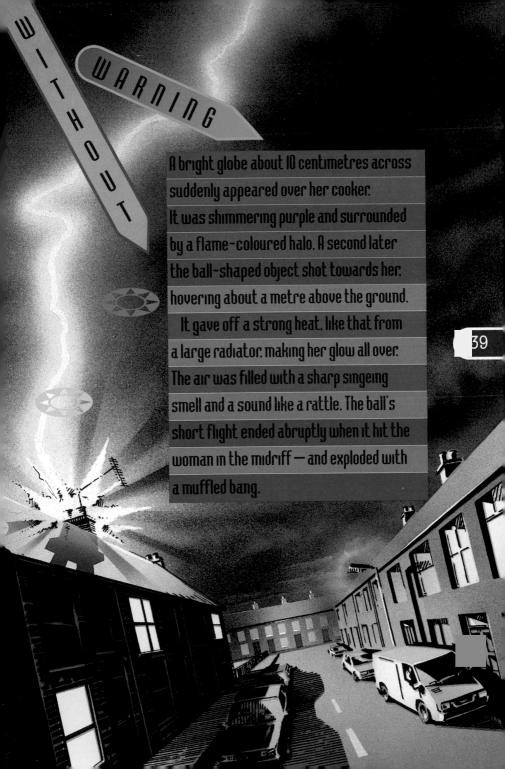

WARNING

WITHOUT

A bright globe about 10 centimetres across suddenly appeared over her cooker. It was shimmering purple and surrounded by a flame-coloured halo. A second later the ball-shaped object shot towards her, hovering about a metre above the ground. It gave off a strong heat, like that from a large radiator, making her glow all over. The air was filled with a sharp singeing smell and a sound like a rattle. The ball's short flight ended abruptly when it hit the woman in the midriff — and exploded with a muffled bang.

39

The ball seemed to hit me below the

belt and I automatically brushed it from

me and it just disappeared. Where I

40

brushed it away there appeared a

redness and swelling on my left hand.

IT SEEMED AS IF MY GOLD

WEDDING RING WAS

BURNING INTO MY FINGER!

When the ball struck her it damaged the woman's dress and tights.
Her legs were not burned, but they did become red and numb. Her vivid
description, and above all her damaged clothing, gave scientists some
of the first hard evidence that ball lightning really does exist...

Physicists at Royal Holloway College in England, were able to examine her ruined clothes. From the size of the holes, and the charring at the edges, they worked out the heat energy needed to burn through the material. For once, the strange fireballs reported over the centuries had left a trail that *could* be investigated.

CHARACTERISTICS OF BALL LIGHTNING

* APPEARS DURING THUNDERSTORMS
* LASTS FROM A FEW SECONDS TO A FEW MINUTES
* SOMETIMES BUZZES OR FIZZES
* SMELLS OF SULPHUR, LIKE A USED FIREWORK
* VANISHES WITH AN EXPLOSION OF LIGHT AND HEAT, OR JUST FADES AWAY

Crazy Balls

41

For hundreds of years people have been amazed and terrified by flying round or oval-shaped fireballs. They have appeared inside buildings, in the open air, down chimneys or in fireplaces — even on board aircraft.

SOME SCIENTISTS BELIEVE THAT BALL LIGHTNING IS REAL . . .

OTHERS DON'T!

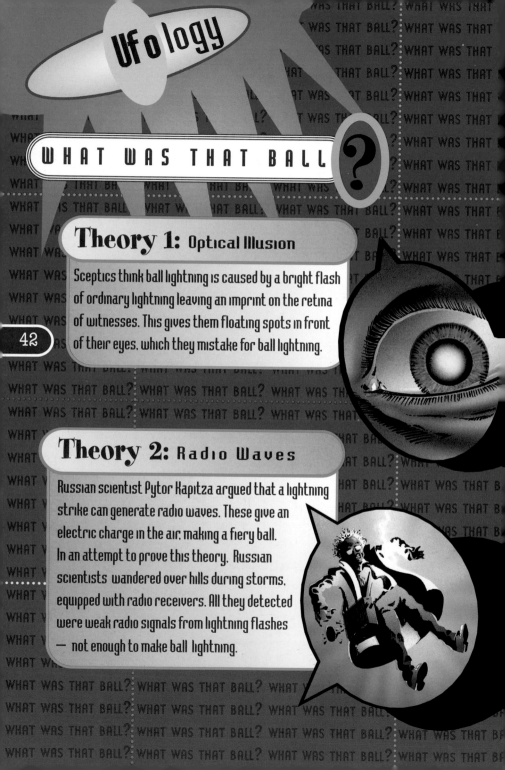

WHAT WAS THAT BALL ?

Theory 1: Optical Illusion

Sceptics think ball lightning is caused by a bright flash of ordinary lightning leaving an imprint on the retina of witnesses. This gives them floating spots in front of their eyes, which they mistake for ball lightning.

42

Theory 2: Radio Waves

Russian scientist Pytor Kapitza argued that a lightning strike can generate radio waves. These give an electric charge in the air, making a fiery ball. In an attempt to prove this theory, Russian scientists wandered over hills during storms, equipped with radio receivers. All they detected were weak radio signals from lightning flashes — not enough to make ball lightning.

Theory 3: Nuclear Ball Lightning

Scientists at the Center for Atmospheric Research, USA, think that thunderstorms could act like natural nuclear reactors. They may make oxygen and nitrogen in the air radioactive, creating high-energy fireballs. However, if this is true then ball lightning should give out deadly doses of radiation. So far no witnesses have died, or shown signs of sickness or radiation burns.

43

Theory 4: Antimatter Ball Lightning

For years cosmologists have guessed that large parts of the universe are made up of antimatter, and when matter and antimatter meet they destroy each other in a huge burst of energy. Scientists at the Atomic Energy Research Establishment in England think ball lightning may be caused when tiny meteorites of antimatter are drawn towards the ground during thunderstorms. They become unstable and explode. A piece of antimatter the size of a pea would be enough to wipe out a whole city.

BOILING HOT BALL LIGHTNING

In 1936 the British newspaper, the *Daily Mail*, published a story of a "red-hot globe", about the size of an orange, which struck a house during a thunderstorm. It cut the telephone wire, burnt the window frame and crash-landed in an 18-litre tub of water. The heat was enough to boil the water.

AIRBORNE BALL LIGHTNING

In March 1963, Eastern Airlines flight number EA 539 was flying between New York and Washington in the USA. Thunderstorms began to toss the plane around and the captain asked the only passenger and the stewardess to keep their seat belts on. A little after midnight the plane was struck by lightning and surrounded by a bright electric glare. In the cabin the passenger looked up with a shock.

Human
COMBUSTION

20cm

A glowing ball about 20 centimetres in diameter was gliding down the aisle. It was blue-white and hovering about knee height over the carpet. It passed harmlessly by him and disappeared from the aircraft near the toilets at the rear.

Scientists like to have reliable witnesses, and in this case they had one. The astonished passenger was Roger Jennison, a professor of electronics at Kent University in England.

There have been reports of people suddenly bursting into flames, for no reason. This phenomenon is called "human combustion". Investigators suspect that they may have been victims of ball lightning.

THE SAUCER SCARE

acting deputy US marshal part-time police officer

aerobatics spectacular flying that includes loops, rolls and tight turns

allies countries that have agreed to help each other in wartime

alternately first one, then the other

emitting letting out, usually smoke or noise

FBI government police force with power to act in any state of the USA

guided missile rockets fired from the ground and aimed at special targets

km/h abbreviation for metric speed (kilometres per hour)

meteorologist scientist who studies the weather

naive simple, easily "taken in"

Soviet Union huge nation of 15 states, dominated by Russia, that broke up in 1990

ufology the study of everything related to UFOs ("unidentified flying objects")

 SIBERIAN DISASTER

asteroid minor planet that's a small body of rock or iron orbiting the Sun

atomic bombs nuclear bombs developed in the 1940s by "splitting the atom" – and capable of terrible far-reaching and long-term destruction

fireball the atmosphere acts as a "shield" for the Earth, burning up small objects before they hit the ground

incinerated burnt to a crisp

meteorite piece of rock or metal from space that reaches the surface of the Earth; most are thought to be fragments of asteroids

radioactivity dangerous waves of nuclear energy

Tungus wandering tribe who follow the herds of reindeer in central Siberia

vaporise dissolve from a solid or liquid into a gas

 CAUGHT ON TAPE

Bermuda Triangle area between Bermuda, the West Indies and the US coast where many ships and aircraft have mysteriously or allegedly vanished over more than a century

calculated worked out with figures

civilian person who is not a member of the armed forces

computer-enhanced images that are improved by using computers

estimate figure decided on the

information available; it may not be accurate

optical physicist scientist who studies light and lenses

orbiting going round a larger object, usually in a circle

radio call sign name used to identify one plane from another when a pilot talks to flight control

watt standard measurement of electric power, named after steam engine inventor James Watt; a kilowatt is 1,000 watts

 ## UFO ALERT!

altimeter instrument for measuring altitude

altitude height when referring to airborne objects, mountains and the weather

humanity the human race and its achievements

interceptor fighters fast fighter planes used to detect and catch invading aircraft

Mach measurement of objects flying at the speed of sound (named after Austrian scientist Ernst Mach); when an aircraft flies faster than Mach 1, it is said to "break the sound barrier", and the fastest jets can reach Mach 3

radar method of detecting on screens the altitude and speed of objects in the air, developed by the British during World War 2; the name comes from **RA**dio **D**irection **A**nd **R**anging

 ## UFO IDENTIFIED

antimatter the opposite of matter, of which all things are composed; scientists can produce antimatter in special machines and believe that it exists naturally – but can't prove it

characteristics main aspects and features of something

combustion burning; "spontaneous combustion" is an explosion without any visible means of ignition

cosmologist scientist who studies the Universe (the cosmos) and how it began

ion electrically charged atom

radiation dangerous loss of energy from a source without physical contact

retina layer of the eye that is sensitive to light and sends images to the brain via the optic nerve

sceptic person who doesn't believe in something, especially where there is little scientific evidence

singeing scorching, or lightly burning

turbulence unstable, gusty patterns of wind

Index